Yuri V. Dunaev
Electricity without charges

Copyright © Yuri V. Dunaev
ISBN-13 978-1530049059

The Summary

Material world such as imagined by the Ether friendly physics is built with three basic mutually bound components or spheres. These are proton matter or proton-sphere that has protons as its main components, electron matter or electron-sphere, and the gaseous ether or elon-sphere. The so called electromagnetic forces that attract or repel interacting objects do not result from interaction of electric charges, but rather from interaction between elements of the proton matter and electrons through the mediation of the gaseous ether. The notion of electric charge as understood by modern physics lacks material entity, and the value, by which electric charge is evaluated, is an equivalent of the mass of electron, the one that in its turn is the equivalent of electron's diametric section area. In this connection the EFP consider electric phenomena as exclusively caused by the existence of electrons and their relation to the proton matter, which can explain such phenomena simpler and clearer without use of such notions as electric charge or electric field that had been introduced to scientific usage mainly through misunderstanding and as substitutes to the material entities of the examined phenomena. Particularistic or more precisely electronic approach to the explanation of electric phenomena permits to explain electric potential as kinetic energy of electrons and electric current as flow of electrons energy between different parcels of electronic cloud.

Contents

Chapter 2: Electric current

Conclusion

Introduction

Material world as conceived by the Ether friendly physics is built with three principal interconnected components or spheres. The most explored of them is the proton matter i.e. that, of which the principal components are protons. Besides protons themselves hereto relate the built with protons nuclei of molecules, atoms, and ions. This sphere could be dubbed protonosphere. Less explored is the electron matter or electronosphere that makes the subject of this book, and the third may be the most important and the least explored is the gaseous ether (elonosphere).According to my views that have much in common with those of ancient theoreticians of ether, who built their models in order to explain the phenomenon of gravitation, particularly Nicolas Fatio de Duillier (1664-1753) and Georges-Louis Le Sage (1724-1803), ether that fills up all the space between elements both of protonosphere and electronosphere, is a kind of pseudo-gas made up with minuscule particles that I having found nothing more convenient dubbed *elons* while the pseudo-gas itself – elonosphere. The use of these terms seems me helpful for further explications. At the same time there would be proper to note that the explication of gravitation with help of the ether's models proposed by the said theoreticians having been manifested unconvincing, the models during centuries remained without any use, though in my mind they might be proper and useful both for the explication of the interaction of electrons between themselves and that of proton matter with electrons.In my view elons are much smaller than all other known particles and they are continuously chaotically moving similarly to molecules of habitual gases.

The speed of the elons chaotic motion is much higher than that of molecules of habitual gases that is known equal to the speed of propagation therein of sound, and is in my view equal to the propagation speed in vacuum (officially recognized vacuum) of electromagnetic oscillations, that is to the speed of light.

Electricity as it is conceived by modern science is indissolubly connected with the notion of electric charge that together with some others e.g. masse, long since represents one of the entangled dogmas of modern physical science.

What is electric charge, what is its physical nature, is it a particle or something else, the modern science does not give any direct answer, and instead of direct answer it only indicates that its porters are electrons, protons, and other particles. As declares the modern science, electric charge has two forms: positive and negative. Electrons and protons carry with them charges of equal value, but the positive charge of proton is exactly opposite to the negative charge of electron. If an object has more protons than electrons it is charged positively; if it has more electrons than protons it is charged negatively; if it has equal amounts of protons and electrons the charges balance each other, and the object becomes electrically neutral.

The notions of positive and negative electric charges are inherited from Benjamin Franklin (1706-1790), who furthermore noticed that opposite charges attract and the same ones repel one another. How strongly they attract or repel further explored Charles Augustine Coulomb (1736-1806), who had formulated the results of his explorations in form of the well-known law of Coulomb, according to which electromagnetic forces attracting or repelling the interacting objects, are proportional to the product of their charges and inversely proportional to the distance between them. As to my mind:

1) The so called electromagnetic forces that attract or repel the interacting objects are not exerted by interaction of electric charges, but rather by that of proton matter elements and electrons through the mediation of gaseous ether;

2) The notion of electric charge as understood by the modern physics lacks material entity and the value, by which it is evaluated is an equivalent of the mass of electron, the mass that is in its turn the equivalent of its diametric section area [1].

In this connection the EFP considers electric phenomena as exclusively caused by existence of electrons and their relation to the proton matter, which can explain them simpler and clearer without use of such notions as electric charge or electric field that had been introduced to scientific usage mainly through misunderstanding and as substitutes to the natural entities of the examined phenomena.

Numerous data collected by modern science and technology on electric phenomena make part, on one side, of the section named electrostatics and, on the other side, of the section that systemizes knowledge connected to electric current. The phenomena brought to these sections have their proper particularities, and while disclosing these particularities from the point of view of the EFP it seems necessary to determine their nature and sources.

It is known that almost every material body made from proton matter keeps, in addition to orbital electrons, an associated with it cloud of electronic gas made up with chaotically moving electrons. The electronic gas as concerns its physical properties, has no distinctions from the well studied by the classical physics molecular gases, and in the same way is subordinated to the known gas laws [2]. Nevertheless, the electronic gas has its own particularities, the most significant of which is its fragmentariness that consists in its allotment to separate clouds.

Each of the clouds can be connected to its own porter, which porter can be any material body, even a droplet of liquid, or separate atom, or molecule. The cloud is kept by its porter with forces of Fatio (today they are called coulomb or electromagnetic forces) that act between the proton matter and electrons, keeping them nearby the proton matter.

It is worth to note that the said fragmentariness is proper not only to electron-sphere but also to ordinary molecular gases, which might be proved by that apart from the atmosphere of the Earth there are those of other planets and their satellites.

In each system composed with an electronic cloud and its porter, the said components are continuously interacting and the relation of their amounts determines the ability of the system to interact with other systems. Every misbalance in a body between the amounts of proton and electron matter is considered today as the presence therein of electric charges of such or another sign, whereas as electrically balanced are considered grounded systems.

Important particularity of electronic gas is the chaotic motion of its electrons, whose speed and therefore energy is another distinctive factor of the above mentioned systems. In different systems the electrons chaotic motion speeds may be different. They can distinguish within the same system, and such distinction may most often be kept artificially against the natural property of electronic gas to equalize its energy, and therefore electrons chaotic motion speed in each part of the same cloud. The electrons chaotic motion speed equalization manifests itself in form of electric current that may be used for satisfaction of different human needs.

Summarizing one can affirm that **the nature of different electric phenomena results from the misbalances of two kinds:**

1) **Misbalance between the amounts of proton and electron matters that may be considered as the source of electric charges, and that makes the subject of electrostatics;**

2) **Misbalance between the electrons chaotic motion energies in different places of the same electronic cloud that is the source of electric currents.**

In Chapter 1 the point will be about such phenomena that result from misbalances of the first kind and make the subject of electrostatics, while the point of the Chapter 2 will be about electric currents.

Bibliography to Introduction

1. Yuri Dunaev, Mass and electric charge as two other hypostases of screening area. Dimensions of electron and ethereal pressure. http://gsjournal.net/Science-Journals/%7B$cat_name%7D/View/4358
2. Yuri Dunaev, Ether Friendly Physics and Gas Laws, http://gsjournal.net/Science-Journals/%7B$cat_name%7D/View/5757

Chapter 1: Electrostatics

1.1 Forces of material bodies electrostatic interaction

The forces mentioned in the title to this Subchapter are scientifically named as electromagnetic, although for such denomination they have no specific reason, being with magnetism in no connection. In my turn, I name the attractive electrostatic forces as forces of Fatio, because he the first had discovered the mechanism of their creation, and the repelling forces as forces of Faraday, because it was he, who the first had demonstrated to the science the cage of Faraday, in which electrons are pushing away from each other under the action of these forces, accumulating themselves on its superior surfaces and leaving the middle of the cage free from any electric effects.

1.2 Forces of Fatio

If to imagine a body as the sole and fixed in space filled with gaseous ether under some pressure P, then continuously and from all the directions it would acquire strikes from elons that would elastically rebound, whereas the body itself due to the central symmetry of these strikes would remain still.

One would observe a different picture if near this body to place another one, the same or somewhat different. In this case each of the bodies would acquire elons' strikes from all the directions except from those shaded or screened by the other one.

As a result, there would appear opposite forces F_1 and F_2 that would press the bodies together, the forces that I had named "forces of Fatio".

Before determining these forces let us realize that if e.g. the bodies are of spherical form, the first of them that has full external surface area $s_1 = 4\pi r_1^2$ will undergo the full pressing force of $p4\pi r_1^2$ that being evenly distributed on the whole surface of the body has a resultant equaling nil, while the other will undergo an analogues pressing force of $p4\pi r_2^2$. Being screened by the other body, the full force pressing the first of them will diminish by a value, proportional to the solid angle filled by the other body in the full solid angle with the vertex in the center of the first body. With an admissible accuracy the filled solid angle can be found as a relation of the screening area of the other body to that of a sphere with radius equal to the distance D between the two bodies

$$\varphi_1 = \frac{\pi r_2^2}{4\pi D^2} = \frac{r_2^2}{4D^2}$$

$$(1.2.1).$$

Then we will obtain $F_1 = \dfrac{p\pi r_1^2 r_2^2}{D^2}$, and using the same procedure for calculating F_2, one will be able to notice that

$$F_2 = F_1 = \frac{p s_1 s_2}{\pi D^2}$$

$$(1.2.2), \text{ where}$$

s_1 and s_2 are the areas of diametric sections (in other words masking or screening areas) of the first and second bodies.

If for instance we are going to examine the interaction between electron and proton in atom of hydrogen, the formula (2) could be edited as follows:

$$F_1 = F_2 = \frac{p s_p s_e}{\pi D^2} = \frac{p \eta s_e^2}{\pi D^2}$$

$$(1.2.2a), \text{ where}$$

s_p and s_e are the cross section areas of proton and electron, and η is the relation of the first to the other. In a case of interaction of a body with n free protons and another body with m free electrons the formula (1.2.2) would be represented as

$$F_1 = F_2 = \frac{p \eta s_e^2 nm}{\pi\, D^2},$$

Or if to designate $\dfrac{p \eta s_e^2}{\pi}$ as k,

$$F_1 = F_2 = k\frac{nm}{D^2}$$

$$(1.2.2b).$$

If to identify the electric charge with the screening area created by electron, the formula (1.2.2b) could be identified as a formulation of the law of Coulomb

$$F = k_e \frac{q_1 q_2}{r^2}$$

(1.2.3), where

q_1 and q_2 are supposed to be electric charges, r – distance between them, and k_e is a coefficient of proportionality. There come to mind that the latter could be represented as $k_e = \dfrac{\rho \eta s_e^2}{\pi}$, and that the attractive Coulomb force is nothing else as the one created by the above described mechanism of Fatio.

1.2 Forces of Faraday

To better understand the nature of forces of Faraday it would be advisable to imagine the way of interaction between gases and proton matter. The most demonstrative and instructive example of such interaction is that of the Earth and the atmospheric air. A commonplace is that the air molecules are retained by the Earth thanks to the forces of its gravitation that attracts each of the molecules, and that thanks to such attraction the air remains under continuous pressure directed, although not precisely, to the center of the Earth. It is also known that the heaviest air molecules try to occupy the lowest of the atmospheric levels, while the lightest e.g. those of hydrogen occupy the highest ones.

If only there would not be the forces of gravitation that create the gravitational pressure attracting air molecules to the Earth, the said molecules would abandon it long ago and run out into the open space, which concerns not only air molecules but also electrons that occupy their places at all the levels of the atmosphere till the ionosphere (50 – 1000 km above the surface level), where their presence is especially conspicuous.

13

As a consequence there seems that the tendency to spread out in space and to lessen the respective partial pressure is proper to all gases including electronic. Resisting to this tendency are external forces, and as concerns electrons, such external forces might be those of Fatio that press them to proton matter including the Earth's surface. As a confirmation, may serve [1.1] that as long as in 1804 P.Erman expressed an opinion that the Earth is charged negatively, the opinion that was experimentally confirmed by Jean Charles Athanase Peltier in 1842.

As it was mentioned earlier, electronic gas is characterized by its fragmentariness that is its allotment to separate clouds, each of which is bound to its proper porter that may be any material body including gouts of liquid, separate atoms, or molecules. The cloud is retained by its porter with forces of Fatio acting between the proton matter and electrons and retaining the lasts in proximity to the proton matter.

The most interesting from my point of view is that in spite of its fragmentariness, there are proves to that the electronic gas behaves as an integral body, and one of such proves is a phenomenon of electrostatic induction, when electrons of two neighboring clouds associated with different electro-insulated one from another objects, during an rapprochement of the lasts, try to run away from each other while rising their amount in the remotest and diminishing their amount in the nearing places of the clouds. Owing to the objects' isolation, there comes to mind that the interaction of the electronic clouds takes place thanks to the intermediateness of ether. Analogues by its physical nature is electrons concentration on the surface of conductors and lack of them in their interior that is observable in conductors with empty core, as well as the electrons absence inside a Faraday cage.

14

The above examples prove that any rapprochement of electronic clouds has to overpower the resistance of certain forces, to which in continuation and development of the just used term "cage of Faraday" there may be proper to give name "forces of Faraday".

If to try to find out analogies to forces of Faraday, one would find them in those forces that create pressures in clouds of molecular gas, the pressures that provoke its running away, if only there were no retaining forces that may, e.g. for molecular gases of the atmosphere, be those of gravitational attraction of the Earth.

1.3 Electronic clouds

As it was said earlier, each material body made from proton matter, i.e. matter with the base of protons, i.e. nuclei of atoms, molecules, ions etc. keeps at least one associated with it cloud of electronic gas composed with chaotically mowing electrons (see fig.1a representing body 1 encircled with envelope (cloud) 2 of electronic gas, where for better demonstrability electrons are designated as red crosses).

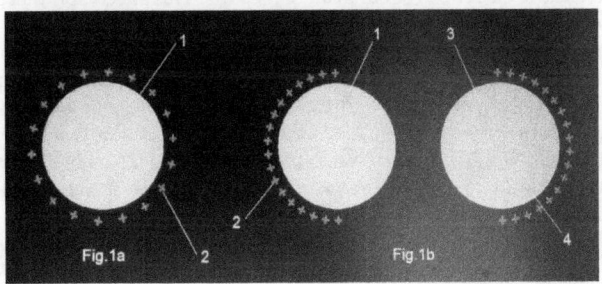

Fig.1a   Fig.1b

The words "at least one" are used here because physical bodies can be made from materials with different electric conductibility. Bodies made from metals that have high electric conductibility have as a rule one electronic cloud for the whole body, the one that envelops it from all sides. On the contrary, bodies made from dielectrics have multitude autonomic and electrically separated micro-clouds that integrally embrace the body, but each of which enveloping only one mini-fragment of the dielectric material. If only there were not the forces of Fatio, that connect electrons with the body's proton matter, the electrons would run away under the action of the electronic gas pressure that passes from electron to electron not only by means of their interaction, but also as a result of the mediation of ether. The last can explain a known fact that in an empty conductor free electrons concentrate themselves on its exterior surface, while in the interior of the body there is no by electrons generated field – the phenomenon that explains the action of Faraday cage. In spite of the apparent autonomy of electronic clouds, they as all other matter present in the ether medium are more or less interconnected, the interconnection becoming stronger as they become closer together. So if to approach the two bodies 1 and 3 of fig.1b at a short distance, their clouds 2 and 4 would shift away as shown on the drawing. The cloud's electrons chaotic motion speed, if only there is no electric current, is equal in all of its parts. This speed and therefore the **kinetic energy of the electrons motion determine its electric potential**. At normal conditions i.e. when the electric potential of a cloud is close to that of clouds connected to ground, the kinetic energy of electrons cannot overcome the forces of Fatio that attach them to the relevant porter, even in those places, in which these forces can be minimal (e.g. in places of surface edges). Nevertheless in cases of substantial increasing of electric potential, said forces may turn out to be too weak and a part of electrons may disconnect from the cloud and jump to nearby air molecules, water droplets, etc. provoking their ionization.

The chaotic motion speed of electrons in to ground connected clouds that one could designate v_0 has to be the same for all of them and serve as a certain point of reference for appreciation of objects electrization. Basing on the widespread opinion that the electric potential is a potential energy [1.2] directly connected with the speed of motion, I assume rational to represent the electric potential of any point of the electronic cloud as

$$p = \frac{\rho}{2}(v - v_0)^2$$

(1.4.1), where

v is the electrons chaotic motion speed in any point of the cloud, and ρ – is its specific mass, and I also assume rational to name this kind of cloud electric potential as "point potential". Then one could determine the electric potential of the whole cloud as

$$P = \frac{M}{2}(v - v_0)^2$$

(1.4.2), where

M is the mass of the whole cloud.

Establishing even a momentary contact between two electronic clouds with different point potentials would generate an electric discharge resulting in creation of a common cloud of **even point potential i.e. one, in which the electrons chaotic motion speeds would be equal.**

As to the potential of individual electrons, by which we could mean their kinetic energies, then if their chaotic motion speed is equal v_0 their electric potentials will be nil. Due to the mediatory action of ether, electronic clouds, close one to another interact repelling each other sometimes together with their porters if only the lasts can freely move (e.g. the leaves of an electroscope), or shift inside the porters, if the lasts cannot freely change their mutual position (as shown on fig.1b). In spite the electronic clouds associated with different porters might seem unconnected; indeed they all make up a common electronic gas subordinated to the same gas laws as our atmosphere.

Therefore if to draw closer two electronic clouds porters, which would be equivalent to diminution of volume and rise of electronic gas pressure, the work of approaching such porters would have to change into the energy of the clouds electrons and rise of their potential. On the contrary, an artificial alienation of electronic clouds has to diminish the energy of their chaotic motion.

At normal conditions the electronic concentration (i.e. the amount of the cloud's electrons in the unity of its volume) in bodies of different materials is different, and as it seems for each of them it has to be constant, higher for denser and lower for less dense materials.

Clouds' electric potentials can be different, but at normal conditions the potentials of all clouds independently of kind of their porter has to be even and equal to the potential of electronic clouds connected with ground i.e. to nil.

1.4 How the body-porter's capability to retain electronic gas depends on its form

Fig.2 imagines a taken for instance body 1 – porter of electronic cloud 2. The body has form of a lens, whose central parcels of the surface are nearest to the whole thickness of the proton matter layers, to begin with the external layer 3 and to end with the nucleus 6. Further from the periphery the distances to the profound layers gradually grow, which diminishes the forces of Fatio attracting electrons of the electronic cloud to the body 1. On the body's ends electrons are pressed to the surface only by the proton matter of the layer 3 and only in much lesser measure by all others.

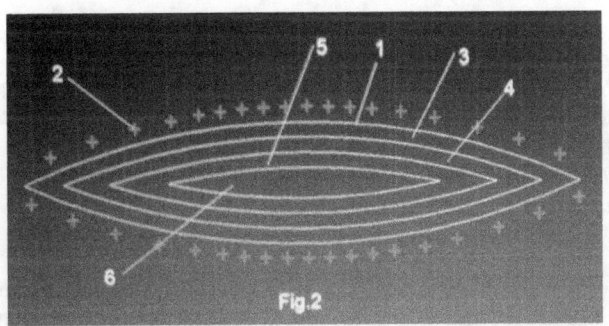

Fig.2

The illustration shows that the pressure made by body-porter on electronic gas depends on its surface curvature. On its prominences and especially on sharpen ends it can be minimal, and on the contrary in its hollows it has to be maximal. It would not be out of place to note that electronic gas even in greater measure than ordinary molecular gases has all the properties of the ideal gas and is subordinate to the gas laws. Therefore if in a certain electronic cloud the electrons chaotic motion energy is even for all its parts, then according to the general gas law the product of the pressure exerted on the cloud by the proton matter and the volume of electronic gas has to be constant. This means that on the porter's sharp ends where the proton matter pressure is minimal the cloud would have the greatest volume, and under certain conditions electrons would jump off the porter's surface till their complete separation.

1.5 <u>Interaction of electrified bodies</u>

The modern electrostatics is based on some dogmas connected with the doctrine of positive and negative charges. According to the dogmas:

- *Positively i.e. "protonly" charged object would apply repelling force to another positively charged object, and quite similarly: negatively i.e. "electronly" charged object would apply repelling force to another negatively charged object.* **Objects with likely charges repel each other;**
- *Positively charged object would attract negatively charged object.* **Oppositely charged objects attract each other;**
- **Any object either charged positively or negatively would attract neutral object.**

In literature one may find out multiple examples confirming the said dogmas. The examples travel from one source to another, but in my opinion not all of them can be estimated convincing. My vision of the electrostatic interaction is based on following:

1) Proton matter of interacting bodies attracts and is attracted by the forces of Fatio, or in other scientifically accepted words by "electromagnetic" forces;
2) Electron matter of interacting bodies repels and is repelled by the forces of diffusion of electronic gas, which forces I for convenience and paying tribute to the great scientist would like to name Faraday forces;
3) If forces of Fatio prevail over Faraday forces the interacting bodies attract each other, and on the contrary if the Faraday forces prevail over the forces of Fatio the bodies repel one another.

In order to determine the said forces let us imagine an electrically grounded body. In spite of the body might be enveloped with an electronic cloud, its electric charge as it was said earlier has to equal nil. If by some means to increase the number of electrons in the cloud, the body will become electron charged, and vice versa, if to decrease this number, the body will become proton charged, the created deficit of electrons exactly corresponding to that proton matter that would be attracted by the forces of Fatio

It may come to one's mind that the here expressed ideas had long ago found their place in the theory of ions' creation. Therefore, in order to show the fallibility of such position I permit myself to note that according to the said scientific theory, ions are created by means of acquiring or loosing orbital electrons, while the proposed hypothesis deals with those free electrons that make up electronic clouds.

The determining of the said forces seems necessary to begin with that of the ratio between the proton and electron matters in both interacting bodies (e.g. in the bodies 1 and 3 of the fig.1b). In the way of example I shall use provisional units of quantity of these matters, units that one could name charges, if only to remember that these new units, used only for example, radically differ from the charges accepted by the official science. Then the proton charge of body 1 could be denoted as p_1, electron charge as e_1, and the respective charges of the body 3 as p_2 and e_2.

Between each unit of proton matter of the bodies 1 and 3 acts their own force of Fatio, and the resulting integral force of attraction has to be proportional to

$P \propto p_1 p_2$ \hspace{2cm} (1.6.1).

Similarly the integral repelling force will be proportional

$E \propto e_1 e_2$ \hspace{2cm} (1.6.2).

Then the conditions of interaction of two bodies will be:

1) $p_1 p_2 = e_1 e_2$ – the bodies do not interact;
2) $p_1 p_2 > e_1 e_2$ – the bodies attract themselves;
3) $p_1 p_2 < e_1 e_2$ – the bodies repel each other.

Taking to account that the both forces have to be inversely proportional to the square of the distance between the bodies D, the formulas (1) and (2) can be defined more accurately

$P \propto \dfrac{p_1 p_2}{D^2}$ \hspace{2cm} (1.6.1a) and

$$E \propto \frac{e_1 e_2}{D^2}$$

$$(1.6.2a),$$

where one might notice a similarity with the Coulomb law.

1.6 Triboelectric effect

As one can get to know from [1.3], triboelectric effect is a type of contact electrification, in which certain materials become electrically charged after they come into frictional contact with a different material. Rubbing glass with fur, or a comb through the hair, can build up triboelectricity. Most everyday static electricity is triboelectric. The polarity and strength of the charges produced differ according to the materials, surface roughness, temperature, strain, and other properties. The triboelectric effect is not very predictable, and only broad generalizations can be made. The most widely known examples of tribo-electrification is charging of amber by friction with wool, which was first recorded by Thales of Miletus who suggested the word "electricity" from the Greek word "electron" that means amber. Other broadly known examples are glass rubbed with silk, and hard rubber rubbed with fur.

In 1757 John Carl Wilcke published a paper describing triboelectric series in which different materials were listed in order of the polarity and charge intensity on separation after they were touched with another object. A material towards the bottom of the series, when touched to a material near the top, will acquire a more negative charge. The farther away two materials are from each other on the series, the greater the charge transferred. Materials near to each other on the series may not exchange any charge, or may even exchange the opposite of what is implied by the list.

Although the prefix "tribo" comes from a Greek word that means "friction", for changing charges two interacting materials need only to contact each other. Explaining the mechanism of such changing, the source [1.3] asserts that after coming into contact, a chemical bond is formed between parts of the two surfaces, called "adhesion", and charges move from one material to the other to equalize their electrochemical potential. This is what creates the net charge imbalance between the objects. When separated, some of the bonded atoms have a tendency to keep extra electrons, and some – a tendency to give them away. Such explanation seems badly grounded, badly comprehensible, and inadequate to the views of the Ether friendly physics on physical nature of electric charges [1.4].

In my view the determining factor of triboelectric effect is the difference between those pressures, with which proton matter retains its electronic clouds. Undoubtedly the retaining pressure has to depend on the object's material, but on account of its creation mainly by its superficial layers it has also to depend on its form and dimensions.

If two electrically disconnected objects 1 and 2 that have electronic clouds with masses m_1 and m_2 are electrically neutral, their unitary masses have energies equal to the energy E of unitary masses connected with ground. According to the main gas law $P_1 V_1 = P_2 V_2 = E$, where P_1 and P_2 are pressures installed in clouds 1 and 2 by the forces of Fatio exerted on clouds by the objects proton matter, and V_1 and V_2 are volumes of clouds unitary masses. The volumes of the clouds themselves have to equal $v_1 = V_1 m_1$ and $v_2 = V_2 m_2$.

If to attach the objects 1 and 2 together and to unite their clouds into the one common cloud 3, the mass of the late will make $m_3 = m_1 + m_2$, and its volume, if only the surface area of the united object is not substantially changed $v_3 = v_1 + v_2$.

The unitary volume of the common cloud will make $V_3 = \dfrac{v_1 + v_2}{m_1 + m_2}$, and based on that the cloud 3 unitary mass energy does not change comparatively with the masses of clouds 1 and 2, $P_s = \dfrac{E}{V_3} = \dfrac{E(m_1 + m_2)}{v_1 + v_2}$.

The mass of the common cloud will evenly distribute itself upon its whole volume, i.e. upon the whole surface of the united body that owe to occur resulting from even electronic gas diffusion, and the mass of electrons enveloping in the cloud 3 the object 1 will make $m_1 = \dfrac{m_3 v_1}{v_1 + v_2}$, and the object 2 – respectively $m_2 = \dfrac{m_3 v_2}{v_1 + v_2}$.

If now to divide back the object 3, the objects 1 and 2 obtained after dividing will have clouds with masses m_1 and m_2, and the differences $m_1 - m_1$ and $m_2 - m_2$ will become measures of the objects electrifying.

Table 1 contains some arbitrary examples of triboelectric objects electrification

Table 1

Nos	m_1	m_2	P_1	P_2	v_1	v_2
			V_1	V_2		
					$m_1 V_1$	$m_2 V_2$
1	10	1	10/0.1	1/1	1	1
2	5	1	5/0.2	1/1	1	1
3	5	5	10/0.1	1/1	0.5	5
4	1	1	10/0.1	1/1	0.1	1

24

Nos	m_1' $\dfrac{(m_1 + m_2)v_1}{v_1 + v_2}$	$m_1' - m_1$	m_2' $\dfrac{(m_1 + m_2)v_2}{v_1 + v_2}$	$m_2' - m_2$
1	5.5	-4.5	5.5	4.5
2	3	- 2	3	2
3	0.9091	-4.0909	9.0901	4.0901
4	0.1818	-0.8182	1.8182	0.8182

Examples 1-4 illustrate electrification of objects composed with materials exerting on their electronic clouds different retaining pressures. Such materials may be those occupying opposite places in triboelectric series. As one may see from Table 1, as a result of the operation of uniting and separating objects, one of them obtains certain amount of electrons, while the other the same amount of them looses. This means that one of them, just that one with greater retaining pressure becomes protonly charged, and the other i.e. that with lesser retaining pressure becomes charged electronly.

Each of the examples 1-4 illustrates interaction of two separate clouds. Existing data nevertheless show that many really existing objects, especially those composed with dielectrics may have hundreds and even thousands of electrically insulated one from another fragments enveloped with their own clouds, such cases occurring the most often. Rubbing a glass stick with a cloth we indeed enter to interaction hundreds or thousands of clouds enveloping the cloth, as well as hundreds or thousands of clouds enveloping the stick.

If for example one of two interacting objects has 1000 clouds in face of 100 clouds of the other one, then in my view in order to simplify the analysis there would seem rational to reduce these amounts to 10:1, and in the analyzed example to multiply the first cloud mass by 10.

The reader might have had an opportunity to work with a metal (the most often steel) screwdriver, while driving in or out small steel screws. He might have noticed that after some minutes of work the screwdriver begins to attract the screws, which may be explained by nothing of magnetism, but by the same triboelectric effect. In small screws the pressures retaining respective electronic clouds are smaller comparatively to those retaining pressures created by the more massive screwdriver. Therefore during the work a part of electrons transfer from the screwdriver to the screws, charging them electronly, while the screwdriver charges protonly.

As a consequence may be that the triboelectric effect can be proper not only to pairs of objects from different materials, mostly those occupying distant places in triboelectric series, but also to pairs of objects from the same material, if only these objects had substantially different dimensions.

1.7 Electrostatic induction and golden leaves electroscope

If to approach an electronly charged body at a sufficiently small distance to electro insulated one, the electrons of the later body under the influence of the forces of Faraday will seek to run away from the charged body, and as a result those parts of the electro insulated body that would be close to the electronly charged one will become protonly charged, while its opposite parts will become electronly charged. It is, as it seems the principle of electrostatic induction, on which is based the work of the golden leaves electroscope shown on fig.3 borrowed from Wikipedia.

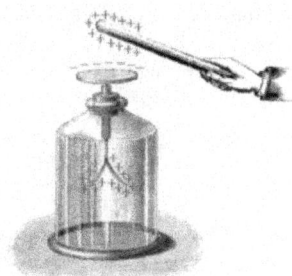

Fig.3

While demonstrating electroscope's operation one approaches a glass stick previously rubbed with woolen fabric to the table of electroscope, whose leaves previously stayed brought together. In my opinion the glass stick having obtained during the electrization electron and not proton charge, transmits it through the induction (i.e. forces of Faraday) to the golden leaves of the electroscope, whose ends diverge under the action of similar forces of Faraday. If though the stick were in some way charged protonly, the electroscopes' leaves would on the contrary clip together by the forces of Fatio, which from the point of view of today conceptions would be qualified as absence of any charge.

1.9 Sourses of static electricity

Prior to description of other and more modern sources of electricity let us have a look on the action of those that similarly to their ancient prototypes base themselves on triboelectric effect and electrostatic induction.

At placed below fig.4 borrowed from Wikipedia there is schematically represented one of the simplest versions of the electrostatic generator proposed in 1929 by the American physicist Van de Graaff. The generator uses in its work the above disclosed triboelectric effect, and is employed either as source of direct electric current or means to obtain electric discharges in form of electric spark.

Van de Graaff Generator

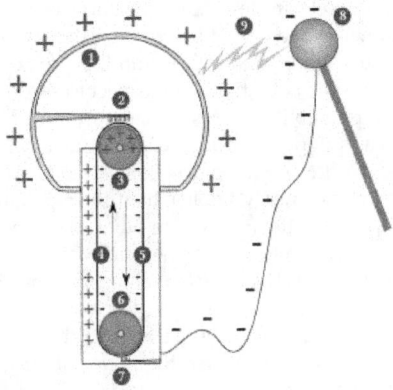

1. hollow metal sphere
2. upper electrode
3. upper roller (for example an acrylic glass)
4. side of the belt with positive charges
5. opposite side of belt with negative charges
6. lower roller (metal)
7. lower electrode (ground)
8. spherical device with a negative charges
9. spark produced by the difference of potentials

Fig.4

The generator has a belt 4,5, fabricated from silk or other flexible dielectric material that runs over two placed one over another metal pulleys 3 and 6, of which the higher is surrounded by terminal 1 in form of a hollow metal sphere, on whose surface accumulates the energy of electrons. Two electrodes 2 and 7 in form of comb-shaped rows of sharp metal points are positioned – one near the bottom of the lower pulley 6, and the other inside the sphere over the upper pulley 3. Comb 2 is connected to the sphere, and comb 7 to the ground. A high DC potential (with respect to earth) is applied to roller 3. As the belt touches the lover comb it receives a charge in form of electrons that cannot hold on the comb's edges and under the forces of Fatio stick to the applied parcel of the belt. When this charged with electrons parcel comes in contact with the upper pulley 3 the electrons acquire kinetic energy i.e. heighten their potential under the influence of the high energetic electrons of the pulley 3. In other words between the pulley 3 and the belt there creates a common electronic cloud, in which in a moment there installs electric current that increases the electric potential (energy) of electrons of the said parcel of the belt up to the potential of electrons of the pulley 3. When this parcel comes into contact with the comb 2 between the parcel and the comb's edges there installs a common electronic cloud, through which the high potential electrons jump to the comb and there from to the terminal 1. Thanks to its dimensions, form, and polished external surface the terminal can accumulate an important amount of electrons that being multiplied by their high electric potential can make up important integral energy e.g. about 450kV. Being continuously in operation the described generator continuously transmits high potential electrons to the terminal 1, and this work continuously compensates losses of electrons' energy, firstly, resulting from terminal's cooling, and secondly, resulting from losses of certain part of electrons due to the corona discharge. As one can understand from the above description the triboelectric effect is used in the described generator, firstly, while the transmission of electrons from the ground through the comb 7 to the dielectric belt 4,5, and secondly while the transmission of the obtained by the belt electrons through the comb 2 to the terminal 1. But if the electrons picked up by the belt in the lower part of the generator have the potential of the ground, those that are transmitted to the terminal have the potential of the higher pulley 3. The last of the potentials if multiplied by the amount of electrons transported to the terminal 1 determines the accumulated therein electric charge power.

Another group of electrostatic generators make induction machines that, as one says today, divide electric charges due to electrostatic induction, and in their work do not depend on rubbing. As a prototype of these machines one can consider the so called electrophorus invented in 1762 by Swedish professor Johan Carl Wilcke and improved by Italian Alessandro Volta.

Schematic representation of the operation of electrophorus, chosen here as an example, is given below at fig.5a-5d.

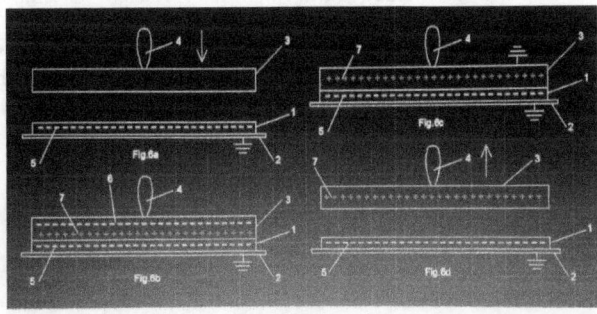

The electrophorus (fig.5a) contains plate 1 made from a dielectric and in the most cases laid on a grounded metal blade or foil 2, and metallic plate 3 with a dielectric handle 4. Before the operation one charges the plate 1 by rubbing it with a piece of fur or fabric, and as a consequence obtains a surplus of electrons designated with little red crosses. Then one applies the electrically neutral plate 3 onto the plate 1 (fig.5b), from which there ensues a redistribution of charges: the electrons 6 of the plate 3 abandon the places nearing the plate 1 and concentrate themselves near its open exterior surface. Having removed for a moment electrons 6 from this open surface, by touching it, for instance, with hand (fig.5c), or by connecting it with a plate of a capacitor (not shown on the drawing) one leaves the proton matter 7 of the plate 3 uncompensated with electrons, and the plate 3 remains charged oppositely to the plate 1. Together with this opposite charge 7 one removes the plate 3 from plate 1 (fig.5d), and if to neutralize the proton charge, one would be able to repeat all the cycle of operations from the beginning.

If to remove the electrons 6 from the plate 3 repeatedly it could be transformed in a generator of electric (electronic) charge and just this principle is employed in electrostatic generators based on electrostatic induction.

On fig.5a-5d there was represented an electrophorus, in which the plate 1 was charged electronly. The below fig.6a-6d show that its operation goes quite similarly if this plate is charged protonly. Then the neutralization of the proton charge of the higher surface of the plate 3 is achieved by flow of electrons from the ground (see fig.6c).

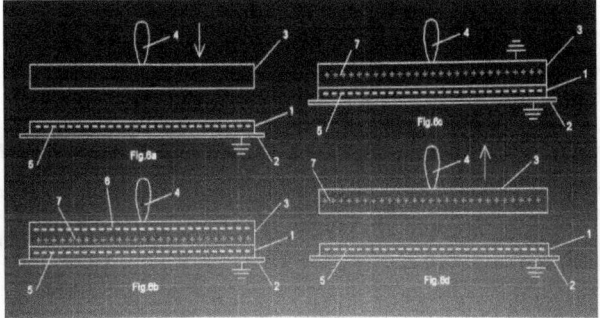

As an example of generators based on electrostatic induction can serve the machine of Wimshurst, the general view of which machine borrowed from Wikipedia is presented below on fig.7.

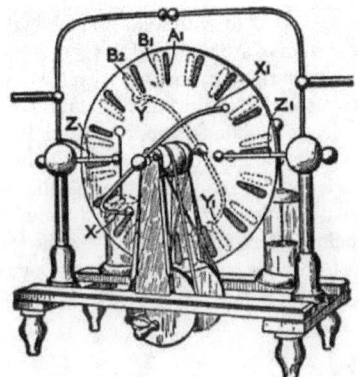

Fig.7

To understand its operation cán help fig.8a-8c.

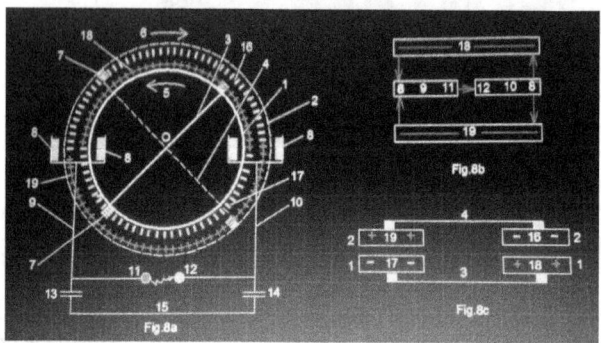

The machine (see fig.8a) has two identical disks 1 and 2, made from electro isolating material, mostly glass rotating on a common axe O in opposite directions designated with arrows 5 and 6. The nearer to us disk 1 is depicted with solid line while the remote disc 2 with dash line. On the periphery of each disk are mounted not represented at the drawing metallic sectors. On the same axe O are rigidly mounted metallic neutralizing planks 3 and 4 with attached to their ends brushes 7 that during the rotation of the disks sweep up the metallic sectors. On two electro isolated supports 9 and 10 there are mounted pairs of brushes 8 that also sweep up the metallic sectors of the both disks and are electrically connected with terminals 11 and 12, made as little metallic balls, and higher relatively to the drawing plates of the capacitors 13 and 14 (made as Leyden jars), the lower plates of the capacitors being interconnected with conductor 15. The planks 3 and 4 are mounted in the plane of drawing and make 90° one to another and 45° to horizon.

The above examined machine in principle does not execute any useful work, because those mechanical efforts applied to rotate the disks 1 and 2 are fully spent to initiate electrons circulation inside the machine itself. The electric discharges created between its terminals are only employed for demonstration of electrostatic effects.

1.10 Capacitors

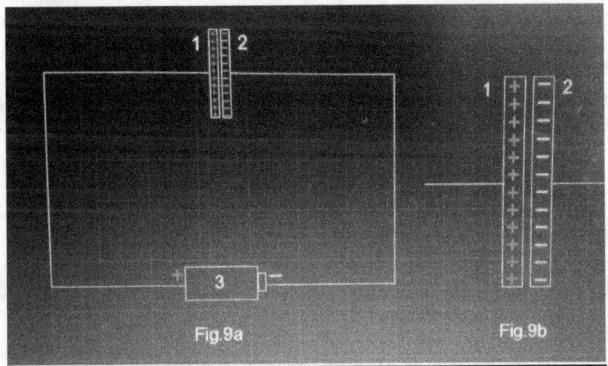

Fig.9a   Fig.9b

In the previous subchapter there was a reference to the capacitors that served to accumulate electrons prior to their use to create spark discharges between the terminals of the Wimshurst machine. In a similar way as the examined static machines, for accumulating static electric charges may serve capacitors.

As it is known a capacitor (see fig.9b) is more often executed in form of two metallic plates e.g. 1 and 2 separated with a thin gap filled with a dielectric that can be vacuum or even air. If to connect the capacitor into an electric schema (fig.9a) with a source of direct current 3, e.g. electric battery, then, as I see it, that of its plates that would be connected to cathode designated unlike to the accepted tradition but similarly to electrodes with sign +, ("cathode" plate 1) would absorb the released electrons up to increasing pressure of the associated with it electronic cloud to some limit level, after which the chemical processes connected with electrons generation would stop. The electrons pressure increasing in the cathode plate 1 would initiate through the mechanisms of induction a flow of electrons off the "anode" plate 2 to the anode, designated with -, where would start the process of electrons absorption. It is clear that the mentioned mechanisms of induction that consist of pushing "anode" electrons away from "cathode" one would be more intensive, lesser would be the gap between the plates.

If after the described process of charging to disconnect the capacitor from the battery and instead of it to connect thereto a resistor, for a moment through the later would flow a current from the charged cathode plate to the uncharged anode plate, and such discharge of the capacitor will be fully analogous to the discharges of the above electrostatic machines.

1.11 Thunderstorms and lightning

Till now we examined electrostatic phenomena generated by human activity. In nature there also exist creatures famous by their use of electrostatic discharges e.g. rays. Nevertheless the most current and spectacular of the natural electrostatic phenomena is lightning.

According to modern views [1.5] lightning are electric discharges occurring during thunderstorms. Thunderstorms arise from a steep rising of warm and humid air. As the warm moist air moves upward, it cools, condenses and forms cumulonimbus clouds that can reach heights of over 20 km. As the rising air reaches its dew point, water droplets and ice crystals form themselves and begin dropping the long distance through the cloud towards the Earth's surface. As the droplets fall, they collide with other droplets and become larger.

Lightning can be of different types. The most common are in-cloud lightning that generate and die away inside the same cloud. More rarely happen lightning occurring between a cloud and ground or between a cloud and there situated objects. Even more rarely happen lightning jumping between two clouds or between ground and a cloud. Eventually the most exotic and in no way explained are ball lightning.

In spite of the nature of lightning has been intriguing mankind all through millenniums, the above exposed concise explanations proposed by science leave a lot of questions. My purpose is basing on the main principles of the Ether Friendly Physics to find out answers at least to the main of them.

In my view the formation of lightning in cumulous clouds occurs in the following way. Under the influence of solar radiation air molecules of upper layers of the cloud and in lesser measure of its lower layers ionize themselves, and the created in this way ionized air molecules while losing external electrons, diminish themselves and by their specific mass become heavier than unionized. Step by step this causes that the positively ionized air molecules go down, and in their place arrive new amounts of unionized air molecules that after reaching the solar light ionize themselves in their turn.

Together with the unionized air there come up molecules of water steam that after reaching a certain height condense creating water droplets.

Electrons freed during the ionization associate themselves (in other words stick), under the action of the forces of Fatio, with the created droplets that leads to their charging, and the associated electrons according to laws of diffusion that demand their maximum scattering, place themselves on external surfaces of the droplets. Under the action of the forces of Fatio the droplets themselves stick together, and the associated with them electrons go over together with them to the newborn agglutinated drops. The electrons concentration is growing with the drops size growing, and a simple calculation shows that such concentration grows proportionally to the growing of the drop's size.

Under the action of the above mentioned diffusion forces or those of Faraday, the electrons of the drop's surface try to run away to all sides, but the forces of Fatio continue to retain them on the drop's surface. Finally there arises a situation when the Faraday forces will provoke the rupture of the drop and the creation in its center of a cavity filled with water steam. The drop disrupted in such manner will transform itself into a spherical bubble, in which the external watery envelop externally covered with a layer of electrons will embrace a cavity with water steam. Evidently such a bubble has all the features of ball lightning, while at the same time it seems evident that the formation of such bubbles is not an extraordinary phenomenon. The truth is that such structures have to be very instable, and the most of them disintegrating inside the cloud are incapable to reach places proper to observation.

In the process of further electrization the bubbles grow due to sticking to neighboring electrified drops, and during such growing the diffusion forces, pushing away superficial electrons, are continuously balanced by the forces of the superficial tension of the bubbles envelops. Comparatively to ordinary drops the descending speeds of such bubbles are substantially lower, and their lowering to the ground becomes poorly probable.

Nevertheless there exist all the reasons to believe that, though extremely rarely the above described watery bubbles can descend to the surface of the earth, because during centuries there were observed and keep to be observed the appearances of the so called ball lightning in form of luminous spheres with diameters from a pea to some meters and with the existence duration from some to some tens of minutes. There exist reasons to consider the glow of the ball lightning as a corona discharge, and that along with its discharging the ball lightning bubble loses its tension and finally blows up. As it can be well understood the explosion of some meters bubble can provoke an important damage and is accompanied with a loud crash.

However, the descent of the above watery bubbles to the ground is a unique phenomenon. Usually such bubbles blow up in the interior of the cloud, and the explosion of one of them initiates a successive explosion of all the neighboring. Such chain explosion of hundreds or even thousands of bubbles is accompanied with thunder, amplified by numerous echoes, and numerous electric discharges that visually merge to lightning, while the freed electrons the most often immediately cling to millions of newly created water droplets. The air after a thunderstorm is always ionized and ozonized.

On the other side, the momentary spraying of the bubbles water can provoke its overcooling and freezing of thousands of thousands droplets that can initiate emerging of hail.

Electrons freed by a thunderstorm can associate themselves not only with water droplets. Under the action of the forces of Fatio they can cling to any objects and even to living beings; and if such objects or beings have sufficiently pointed prominences, as for instance in lightning conductors, towers, masts, funnels, aircraft wings, animal fur fibers, etc., such associated with objects and beings electrons can lift of them off analogously to corona discharges of electric power lines. Such lifting off of electrons is accompanied by glows, crackles, and hissing, characteristic to the long ago known, but till now unexplained saint Elmo's fires.

Bibliography to Chapter 1

1.1 https://en.wikipedia.org/w/index.php?title=Atmospheric_electricity&oldid=6
91411176
1.2 https://en
wikipedia.org/w/index.php?title=Electric_potential&oldid=671272078
1.3 https://en.wikipedia.org/w/index.php?title=Triboelectric_effect&oldid=687534957
1.4 Yuri Dunaev, Mass and electric charge as two other hypostases of screening area. Dimensions of electron and ethereal pressure
http://gsjournal.net/Science-Journals/%7B$cat_name%7D/View/4358
1.5 http://en.wikipedia.org/w/index.php?title=Atmospheric_electricity&oldid=6640798
40

Chapter 2: Electric Current

2.1.Preface

As declare modern science, electric current is that of electric charges. In electric circuits such charges are most commonly transported by moving electrons of conductors. In electrolytes they can be transported by ions and in plasma by both ions and electrons.

In my article [2.1] in conformity with the proved by the Ether friendly physics affirmation about fallibility of the accepted by modern science concept of electric charge, there was declared that electric current in metallic conductor is not that of electric charges or their porters – electrons, but rather – of their kinetic energy.

This opinion bases itself on that the electric current conductors e.g. resistors contain electronic gas in form of electronic cloud with electrons that while moving chaotically are trying to run with an equal velocity, thanks to which if in some place of the cloud they run for a while with a different velocity, the next moment their velocities and their kinetic energies would get equal to the all others. Although their speeds are incomparable, electricity conduction has a long ago noticed analogy with conduction of heat, which is easy to understand if to compare masses of electrons and atoms or molecules of heat conducting objects. Electric energy is transported from electron to electron, the one and the other practically do not changing their place in the cloud, as well as heat is transported from molecule to molecule that also do not change their place in the object. As electron's electric potential there figures in my opinion and it will be proved later, the energy of its chaotic motion, as well as by analogy heat potential of a molecule or its temperature would be the velocity of its rotation [2.2].

In the same way as heat conduction motor is the temperature difference at the ends of a heat conducting object, the electric current motor is in my opinion the electric potentials difference at the terminals of resistor.

2.2 Example of electric circuit

At fig.10 as the simplest example there is imagined a schema borrowed from Wikipedia of a closed electric circuit composed with a source v of direct electric current and a resistor R. As one may imagine, the elements of the circuit, especially resistor, contain an electronic cloud that has in its lower relatively to the drawing part potential p_1 and in its higher part – potentialp_2.

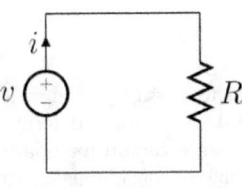

Fig.10

In conformity to the accepted concept and in harmony with the mass conservation principle that might be better to be named "matter conservation principle", the cloud electrons number has to remain unchanged, because it has neither source of their income nor possible losses, and this concerns not only the cloud as a whole but each of its fragments contained in the source of DC, resistor, and connective conductors.

The potentials difference $p_1 - p_2$ is kept by the work continuously accomplished by the source of current and continuously wasted in the resistor while transforming itself to heat.

2.3 Law of Ohm and First law of Joule

Modern views on electricity have as their base two widely known laws, the connection between which, if to base oneself on those views on electric current that is now accepted by modern science is hardly understandable. These two laws are: the law of Ohm and the first law of Joule, or in other words – the law of Joule – Lenz.

The first of the above laws, if applied to the above schema would declare that the current I going through the resistor equals the tension$U$, that according to accepted views is a dropping of potential on the resistor's length, divided by its resistance$R$ in conformity to the formula

$$I = \frac{U}{R}$$

(2.3.1),

that can also be interpreted as

$$R = \frac{U}{I}$$

(2.3.2).

Electric resistance of a conductor is its opposition to passage through it of electric current. As reports the Wikipedia the resistance of a conductor is determined by two factors: by its material and form. In a conductor from a certain material that for the simplicity of further calculations would be appropriate to consider as a segment of the length l and of the same cross-section area A the resistance will be proportional to the length and inversely proportional to the cross-section area that might be represented by formula

$$R = \rho \frac{l}{A} = \frac{l}{\sigma A}$$

(2.3.3),

in which ρ is an inherent to the material resistivity that is a value inverse to the electro conductivityσ.

The equations (2.3.2) and (2.3.3) give together

$$\frac{U}{I} = \frac{\rho l}{A}$$

(2.3.4).

The equation (2.3.4) allows for imagining the tension in form of a product

$$U = \alpha \rho l$$

(2.3.5),

and the current – in form of another product

44

$I = \alpha A$ $\qquad\qquad\qquad$ (2.3.6).

According to **the second of the above mentioned laws** (the first law of Joule) during flowing of electric current through a conductor therein takes place generation of heat H according to formula

$H = I^2 R = UI$ $\qquad\qquad\qquad$ (2.3.7).

The heat in my view is being generated as a result of transferring to it of the electrons chaotic motion energy, and with regard to the formulas (2.3.5) and (2.3.6) one may obtain

$H = \alpha^2 \rho l A$ $\qquad\qquad\qquad$ (2.3.8).

The product lA represents the conductor's volume; therefore in order that H represents energy it would be necessary that $\alpha^2 \rho$ be that fraction of energy that would belong to a unit of this volume.

One could resolve the problem if to imagine, firstly, that α is nothing else than the drop of the electrons chaotic motion velocity from the beginning to the end of the conductor

$\alpha = \Delta v = v_1 - v_2$ $\qquad\qquad\qquad$ (2.3.9),

and secondly, that ρ is proportional to half a mass of the electrons contained in the conductor's electronic cloud unitary volume; and here the half is chosen because electrons of the cloud lose their velocities progressively in proportion to the distance from the incoming terminal. Then $\rho l A$ will represent half the mass of the conductor's cloud electrons.

$\rho l A = \dfrac{M}{2}$ $\qquad\qquad\qquad$ (2.3.10),

and resistivity will make

$\rho = \dfrac{M}{2lA}$ $\qquad\qquad\qquad$ (2.3.11).

Then H will determine that quantity of energy which has been generated as heat

$H = \dfrac{M}{2}\alpha^2 = \dfrac{M}{2}(\Delta v)^2$ $\qquad\qquad\qquad$ (2.3.12),

which does completely agree with the generally accepted notion of kinetic energy.

45

2.4 Determining key electric values

Resolving together the equations (2.3.3) and (2.3.11) one obtains:

$$R = \rho\frac{l}{A} = \frac{M}{2lAA}\frac{l}{A} = \frac{M}{2A^2}$$
(2.4.1).

Resolving together the equations (2.3.5), (2.3.9), and (2.3.11) one obtains:

$$U = \alpha\rho l = (v_1 - v_2)\frac{M}{2lA}\, l = (v_1 - v_2)\frac{M}{2A}$$ (2.4.2).

And resolving together the equations (2.3.6) and (2.3.9) one also obtains

$$I = \alpha A = (v_1 - v_2)\, A$$
(2.4.3).

With regard to the equation (2.4.2) it is easy to conclude that the electric potential in a point of a conductor's electronic cloud has to equal

$$p = \frac{vM}{2A}$$
(2.4.4),

where v designates the electrons chaotic motion velocity in the chosen point.

Here it is worth to notice that the ideas expressed in this subchapter have very much in common with those expressed earlier in the subchapter 1.4, especially when to compare the formulas 1.4.1 and 1.4.2 with the above formulas 2.4.2 and 2.4.4.

Verification

As verification one may compose:

- the equation of the Ohm law: $I = \dfrac{U}{R}.$ $(v_1 - v_2)\, A = (v_1 - v_2)\dfrac{M}{2A} \cdot \dfrac{2A^2}{M}$,

- the equation of the First Joule law: $H = UI.$ $(v_1 - v_2)\, A \cdot (v_1 - v_2)\dfrac{M}{2A} = \dfrac{M(v_1 - v_2)^2}{2}$.

The accomplished substitutions to the Ohm law convert it to identity, and those to the First Joule law do it to the traditional formula for kinetic energy, which confirms the correctness of the expressed ideas and the organic unity of the both laws.

2.5 Alternating current

According to modern views alternating current is such one, in which the electrons' stream direction is continuously changing. It means that if for a moment it went from point A to point B, the next moment it would go from B to A. According to EFP, alternating electric current between connected points of a certain electric circuit is characteristic in that if at a moment electrons of the point A have the potential P_1 and electrons of the point B – the potential P_2, the next moment the electrons of the point A would have the potential P_2, and the electrons of the point B – the potential P_1.

2.6 Mechanisms of direct current creation.

The main attention of the previous subchapters was focused on explanation of physical nature of the main electric values and problems related with transformation of electric energy to heat. They did not though contain information how in the schema represented at fig.10 there is generated and maintained the electric current used for obtaining heat in the resistor according to the First law of Joule.

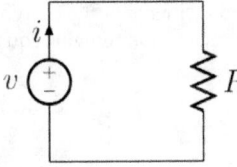

Fig.10 (copy)

The previous subchapters did not disclose the mechanisms of direct current generation and their particularities depending on chosen type of DC-source. Therefore the aim of this subchapter is to disclose the way by which in a DC-source there is generated the energy thereafter used in electric schemas, in particular for obtaining heat.

Today the most widely used of DC-sources are electric batteries – the descendants of those voltaic batteries that had helped to make the plurality of the outstanding electric discoveries of the 19th century. With no special attention to the electrochemical subtleties of their operation, there would be sufficient to indicate that it will result with emerging of electrons on one of the terminals (cathode, marked on the schema with -) and absorption of electrons on the other (anode, marked on the schema with +). The current direction is marked here with an arrow that in accordance with a generally accepted agreement is opposite to the direction of the supposed movement of electrons. The indicated result of battery's operation suggested and keeps suggesting to scientists that the same electrons that emerge on the cathode go through all the chain and absorb themselves on the anode, and that their transfer from terminal to terminal does make up electric current.

As it was already indicated, the schema's elements, particularly the resistor and connecting conductors contain an electronic cloud made up with chaotically moving electrons, and before the connection to the mentioned elements of the current source in form of a galvanic battery the speed of electrons chaotic motion remains equal in all the regions of the cloud.

After the battery's connection its cathode starts to produce electrons that while inserting in the cloud in the place of its connection to the cathode raise there its pressure and therewith – the energy of electrons chaotic motion. In the meantime the anode while absorbing electrons in the place of its connection to the electronic cloud lowers there the electrons pressure together with their energy and speed of their chaotic motion.

The created in this manner energies' difference, as well as that of the cloud's electrons chaotic motion speeds near the galvanic battery terminals become therefore the driving force of electric current, whose particularities are disclosed in the previous subchapters.

Other widespread DC-source is dynamo, in which the driving energy surplus of the electronic cloud electrons is continuously elaborated in the machine rotor armature thanks to the continuous motion of its windings relatively to the magnetic fields created by the stator permanent or electro-magnets.

It would not be out of place to have a look on one more type of generators, that which would be difficult to associate with DC-generators, because the current they provide is extremely transitory and quickly falls to null. These are so called electrostatic generators, or electrostatic machines that produce charges of very high potentials. These high-potential charges created at one of the machine's terminal are actually electronic clouds of high concentration, the one that is continuously raised thanks to mechanical energy applied through tribo-electric effect or electrostatic induction. The momentary discharge of such cloud forms short living electric current mostly in form of spark. Electrostatic generators started to be developed and used for studying electric phenomena yet in 18th century, and their use in this field keeps going up to the present time. They have not nevertheless any industrial or residential applications.

2.7 Mechanisms of alternating current creation

Here above there was a talk about dynamo-machines in which the driving energy of electronic cloud is continuously produced in the machine's rotor armature thanks to the continuous motion of its windings relative to the magnetic fields created by permanent or electro-magnets of the stator. In AC-generators the direction of the relative motion of the rotor's armature and stator's magnetic fields is continuously changing, which results in that electrons of the electronic cloud obtain drawing impulses of changing directions.

The energy of electrons chaotic motion on the rotor's windings terminals is continuously changing, and therefore is changing the energy transfer direction in the connected thereto electric schemas.

2.8 Capacity of capacitors

Although the capacitors are in no way related to the DC-sources, their operation analysis might create an opportunity to confirm the ideas expressed in the previous subchapters.

As it is known a capacitor is most likely realized in form of two metallic plates, separated by a thin gap filled with dielectric that may be vacuum or even air. If to insert the capacitor into an electric schema with a DC-source, e.g. an electric battery, then, as we see, that of its plates, which is connected to the cathode ("cathode" plate) will absorb electrons emerged from the cathode up to the elevation of pressure in the associated with it electronic cloud to a certain limit, after which elevation the chemical processes associated with electrons generation will stop.

The electrons pressure elevation in the cathode plate, by means of induction mechanisms will initiate the electrons drive from the "anode" plate to the anode, where the process of electrons absorption will start. It seems to be clear that the said mechanisms of induction that push the anode electrons away from the cathode ones would be as more intensive as smaller would be the gap between the plates.

If after the end of the described process of the capacitor's charging, to disconnect it from the battery and to replace it with the resistor, just for a moment a current would flow from the charged cathode plate to the uncharged anode one.

In literature [2.3] one may find information about the so called capacitor's capacity that might signify its suitability to accumulate electric charge. According to the said source, capacity is pertinent to all charged objects, but the most widespread means for conservation of energy is parallel plate capacitor. In a parallel plate capacitor the capacitance is directly proportional to the surface area of its conductive plates and inversely proportional to the distance between them. If the plates' charges make respectively $+q$ and $-q$, and V represents the electric tension existing between the plates, the capacity may be calculated according to formula

$$c = \frac{q}{V}$$

(2.8.1),

from which one could not conclude that capacity must be directly proportional to the surface area of its conductive plates and inversely proportional to the distance between them. Any connection between the formula (2.8.1) and relations between capacitor's dimensions being unseen in other places of the said source, the finding of such connection using the relations found in the previous subchapters would become a good opportunity for their confirmation.

First of all let it be clear that according to our convictions in a charged capacitor the cathode plate electric charge represents an electronic cloud, with electrons possessing certain energy and therefore a certain electric potential. On the contrary, the anode plate seems to be exempt from electrons and its electric potential makes null. Therefore the electric tension between the plates in accordance to the formula (2.4.2) has to equal

$$V = v\rho h$$ (2.8.2), where

v is the cathode plate electrons chaotic motion speed, ρ is the specific mass of the electronic cloud associated with the cathode plate, and h is the dimension of the gap between the plates.

According to formula (2.3.11) $\rho = \dfrac{M}{2lA}$, where M is the electronic cloud's mass i.e. its charge (marked in formula (2.8.1) as q), l – is its length (in our case – its width), and A – is the area of its section perpendicular to the length (i.e. in our case just of its surface).

Then $V = v\rho h = vh\dfrac{q}{2lA}$, and the capacity has to be

$$C = \frac{q}{V} = \frac{2lA}{vh} = \frac{2l}{v} \cdot \frac{A}{h}$$ (2.8.3).

Formula (2.8.3) is an indubitable proof of correctness of the affirmation that the capacity of two plate capacitor is directly proportional to its plates' surface area and inversely proportional to the distance between them. It is also a proof of correctness of the principles expressed in subchapters 2.3 and 2.4. As regards the electrons chaotic motion speed in the cathode plate and width of the electronic cloud, these items have to depend on the characteristics of the electric schema and particularly of the source of tension.

Bibliography to Chapter 2

1.1 Yuri Dunaev, Elecric current in Ether friendly physics
http://gsjournal.net/Science-Journals/%7B$cat_name%7D/View/6222
1.2 Yuri Dunaev, Heat, Temperature, and Mechanism of Heat Conduction
http://gsjournal.net/Science-Journals/%7B$cat_name%7D/View/5954
https://en.wikipedia.org/wiki/Capacitance

Conclusion

Material world, and the electric phenomena undoubtedly make one of its numerous facets, is built from material particles. Particles exist separately or linked with other particles, making up certain structures, but in such or other way both separate particles and material structures keep motions that are distinguished by speed i.e. energy. In nature existing no other basic characteristics of matter beside its amount and speed, in the context of this book there seems unnecessary for understanding physical phenomena the notions of electric charge, as well as electric or electromagnetic field. The proposed book is destined to show that the electric phenomena can be explained simpler and more understandably without using these notions introduced to scientific usage mainly through misunderstanding and as substitutes to the natural causes of the examined phenomena.

The book cannot pretend to fully disclose all the electric phenomena. It is destined only to show that all of them examined or not, can be explained by interactions of electrons with proton matter, by those of electrons with ether that I also imagine composed with material particles, or by those of electrons between themselves.

The book cannot examine and give the "electronic" explanation to all the 21st century electrotechnics. Nevertheless there is a hope that with help of enthusiasts of EFP there will stand a new electrotechnics exempt of unnecessary dogmas and excessive complications.

www.ingramcontent.com/pod-product-compliance
Lightning Source LLC
Chambersburg PA
CBHW070407190526
45169CB00003B/1144